Forestry Commission

Field Book 14

Herbicides for Farm Woodlands and Short Rotation Coppice

by Ian Willoughby

Forestry Commission Research Division

and David Clay

Avon Vegetation Research
P.O. Box 1033
Nailsea
Bristol
BS19 2FH

LONDON: HMSO

ISBN 0 11 710336 5
FDC 414:441:236.1:307

Keywords: *Herbicides, Farm forestry*

Abstract

This publication gives information about the use of herbicides in farm woodlands and short rotation coppice. Recommendations are given for suitable herbicides for a range of crop and weed species.

Disclaimer

Enquiries relating to this publication should be addressed to:

The Research Communications Officer
Forestry Commission Research Division
Alice Holt Lodge, Wrecclesham
Farnham, Surrey GU10 4LH

Contents

1 Introduction

Managers involved in the establishment of new woodland and short-rotation coppice on fertile lowland sites face distinctive vegetation management problems. Ex-agricultural and improved grassland sites are normally much more fertile than the unimproved grassland sites that have been the subject of forestry planting for most of this century. This fertility is associated with a large weed seed bank, which can result in prolific weed growth once rotational agriculture ceases. These weeds compete strongly for nutrients, light and moisture, and may lead to stunted tree growth or complete failure of new plantings. Farm forestry, or the establishment of farm woodlands, here refers to new planting on ex-arable or improved grassland sites, as defined in the Woodland Grant Scheme III.

Herbicide treatments developed for use in conventional forestry situations may need to be repeatedly applied to control weeds in farm woodlands, and often need to be directed away from crop trees to prevent damage. Farmers may require less labour intensive, more mechanised methods.

In order to address these problems, the Forestry Commission Research Division has a continuing screening programme to identify herbicides that will give effective weed control on arable and improved grassland sites, while being safe for use over broadleaf and coniferous tree species.

Recent off-label approvals have widened the range of products which can be used in farm forestry, particularly in short rotation coppice. This Field Book gives guidance primarily on products with specific approval for use in farm woodlands, that are safe to spray over trees, and will control agricultural weeds. Users are advised to refer to Forestry Commission Field Book 8 *The use of herbicides in the forest* (1995 edition) for more detailed guidance on broad spectrum contact herbicides with forestry approval, such as glyphosate, glufosinate ammonium, triclopyr, etc., that it may be necessary to use in addition to the more selective products listed here. Field Book 8 also provides essential information on safety, good working practices and application equipment.

2 Approval status

Under the Control of Pesticides Regulations 1986, all pesticides, including herbicides, used in farm woodlands must be approved by the Pesticides Safety Directorate of MAFF for that use. Approval may be full, with a label recommendation, or may be off-label as explained below.

Full approval

Products with full on-label or specific off-label approval for use in forestry, **CAN** be used in farm woodlands and short rotation coppice. The reverse **DOES NOT** apply.

The only herbicides with approval for use in forestry that are treated in detail in this publication are atrazine (Atlas Atrazine/Unicrop Flowable Atrazine), isoxaben (Gallery/ Flexidor) and propyzamide (Kerb) because they are useful in mixes with other products in farm woodland situations. Glyphosate and glufosinate ammonium are contact herbicides that are included for comparison because of their broad spectrum of activity.

Propaquizafop (Shogun/Falcon) has full-label approval for use in farm forestry.

Specific off-label approval

In some instances the Pesticides Safety Directorate will issue specific off-label approval for existing products, which may be of importance for a minor use such as forestry, but where anticipated sales volumes are not sufficient to persuade manufacturers to carry out the research and development required to obtain full-label approval. In these cases, all applications are made at the user's own risk, and all conditions of use detailed on the product label must still be complied with. In addition, users must obtain a copy of the off-label approval document itself, and comply with all conditions of use therein. Copies of all specific off-label approvals referred to in this Field

Book are included in the Appendix. Specific off-label approvals relate to individual products, not to active ingredients. It is not permissible to substitute an alternative product with the same active ingredient.

Products with specific off-label approval:

Clopyralid (Dow Shield) – specific off-label approval for use in forestry.

Metazachlor (Butisan S), cyanazine (Fortrol), fluazifop-p-butyl (Fusilade 250EW) and pendimethalin (Stomp) – specific off-label approval for use in farm forestry.

Atrazine (Unicrop Flowable Atrazine) has a specific off-label approval extending its use to broadleaved trees – the full label approval is only for coniferous trees.

Long-term off-label arrangements

In addition to the two main types of approval, namely full on-label and specific off-label approval, certain fields of use may be covered by long-term off-label arrangements, which are valid until 31 December 1999.

The long-term arrangements grant off-label approval to certain fields of use, rather than to specific products. The same basic principles as specific off-label approval apply, namely users must comply with all label conditions of use as well as additional off-label restrictions, and all applications are made at the user's own risk.

Fields of use

The following extensions of use are permitted under the long-term arrangements:

- Herbicides with full or provisional label approval for use on cereals, may be used in the first five years of establishment of new farm woodlands (including short rotation energy coppice), on land previously under arable cultivation or improved grassland (as defined in the Woodland Grant Scheme).

- Herbicides with full or provisional label approval for use on cereals, oil-seed rape, sugar beet, potatoes, peas and beans, may be used in the first year of regrowth following cutting in short rotation energy coppice, on land previously under cultivation or improved grassland (as defined in the Woodland Grant Scheme).

Conditions of use

As well as the usual good working practices required of users, certain **additional** conditions **MUST** be complied with when applying pesticides under the long-term off-label arrangements – these are detailed in the appendix.

Practical implications

The long-term off-label arrangements should allow a wider range of products to be used in the initial years after planting. Farm managers may be familiar with many of them and already using the same products over different agricultural crops. However, the Forestry Commission Research Division can only offer guidance on those products that have been found to be effective and safe to trees, in small-scale trials programmes. Consequently, only those herbicides for which the Forestry Commission has made additional specific off-label applications are detailed in this publication.

The following additional products may be of use in short rotation coppice situations only: cycloxidim (Laser), lenacil (Venzar), metamitron (Goltix WG), napropamide (Devrinol) and simazine (Unicrop Flowable Simazine). Amitrole (Weedazol) may be of use in farm forestry and short rotation coppice situations. Table 1 summarises the approved products, the approval status, mode of action, method and rate of application.

3 Approved products

3

Product mode of action

The products detailed in Table 1 may be divided into three main modes of action:

a. Residual herbicides

Pendimethalin, lenacil, isoxaben, napropamide and simazine are most active on weeds pre-emergence, and have very little if any activity on established weeds.

b. Residual/foliar acting herbicides

Metamitron, atrazine, cyanazine, metazachlor and propyzamide are mainly pre-emergent herbicides, but they do have activity on some weeds post-emergence, either through foliar or root uptake. Only propyzamide is likely to control mature weeds effectively – see weed susceptibility, Table 5, for maximum growth stages of weeds that can be controlled.

c. Foliar acting herbicides

Amitrole, clopyralid, cycloxidim, fluazifop-p-butyl, glufosinate ammonium, glyphosate and propaquizafop are foliar acting herbicides which are applied to emerged weeds. They must be applied at the correct stage of growth to obtain maximum effect (see Table 5), and are unlikely to have any significant residual or pre-emergent effect.

Methods of application

All the products in Table 1 may be applied through hand-held or mechanised applicators, except for fluazifop-p-butyl (Fusilade), which may **ONLY** be applied through mechanised sprayers.

Hand-held applications are most appropriate when trees are unguarded, when directed sprays avoiding shoots are required to avoid crop damage. Mechanised applications are likely to be cheaper, and it may be possible to make use of existing

Table 1 Summary of approved products and their uses

Active ingredient (amount in product)	Product	Manufac.	Field of use				Approval status	Mode of action	Method of application	Product rate	Maximum number of applications per year
			Forestry	Farm forestry	Farm forestry years 1–5	Short rotation coppice after cutting					
Amitrole (225g/l)	Weedazol TL	Bayer			✓	✓	LTOA – farm forestry	Foliar	Hand-held and mechanised	20.0 l/ha	–
Atrazine (500g/l)	Unicrop Flowable Atrazine***	Unicrop	✓	✓	✓	✓	Forestry label approval Off-label approval for broadleaved trees	Residual pre-emergent/ foliar	Hand-held and mechanised	5–13.5 l/ha	Max total to be applied 13.5 l/ha per year
Clopyralid (200g/l)	Dow Shield	Dow-Elanco	✓	✓	✓	✓	Forestry off-label	Foliar	Hand-held and mechanised	1.0 l/ha	2
Cyanazine (500g/l)	Fortrol	Cyanamid		✓	✓	✓	Farm forestry off-label	Residual pre-emergent/ foliar	Hand-held and mechanised	4.0 l/ha	1
Cycloxidim (200g/l)	Laser **	BASF				✓	LTOA – Short rotation coppice	Foliar	Hand-held and mechanised	2.25 l/ha	2
Fluazifop-p-butyl (250g/l)	Fusilade 250EW	Zeneca		✓	✓	✓	Farm forestry off-label	Foliar	Mechanised only	1.5 l/ha	2

Table 1 Summary of approved products and their uses – *(contd.)*

Active ingredient (amount in product)	Product	Manufac.	Field of use: Forestry	Field of use: Farm forestry	Field of use: Farm forestry years 1–5	Field of use: Short rotation coppice after cutting	Approval status	Mode of action	Method of application	Product rate	Maximum number of applications per year
Glufosinate ammonium (150g/l)	Challenge/ Harvest	Hoechst/ AgroEvo	✓	✓	✓	✓	Forestry label approval	Foliar	Hand-held and mechanised	3.0–5.0 l/ha	–
Glyphosate (360g/l)	Roundup Pro Biactive *	Monsanto*	✓	✓	✓	✓	Forestry label approval	Foliar	Hand-held and mechanised	1.5–5.0 l/ha	–
Isoxaben (125g/l)	Gallery 125/ Flexidor 125	Dow Elanco	✓	✓	✓	✓	Forestry label approval	Residual pre-emergent	Hand-held and mechanised	2.0 l/ha	2
Lenacil (440g/l)	Venzar Flowable/ Vizor**	DuPont				✓	LTOA– Short rotation coppice	Residual pre-emergent	Hand-held and mechanised	4.0–5.0 l/ha 1.1–2.0 kg/ha	1
Metamitron (70%w/w)	Goltix WG**	Bayer				✓	LTOA – Short rotation coppice	Residual pre-emergent/ foliar	Hand-held and mechanised	5.0 kg/ha	1
Metazachlor (500g/l)	Butisan S	BASF		✓	✓	✓	Farm forestry off-label	Residual pre-emergent/ foliar	Hand-held and mechanised	2.5 l/ha	3

Table 1 Summary of approved products and their uses – *(contd.)*

Active ingredient (amount in product)	Product	Manufac.	Field of use				Approval status	Mode of action	Method of application	Product rate	Maximum number of applications per year
			Forestry	Farm forestry	Farm forestry years 1–5	Short rotation coppice after cutting					
Napro-pamide (450g/l)	Devrinol	Rhone Poulenc Agriculture				✓	LTOA – Short rotation coppice	Residual pre-emergent	Hand-held and mechanised	2.1 l/ha	1
Pendimeth-alin (400g/l)	Cyanamid Stomp 400SC			✓	✓	✓	Farm forestry off-label	Residual pre-emergent	Hand-held and mechanised	5.0 l/ha	1
Propa-quizafop (100g/l)	Falcon Shogun 100EC	Cyanamid Ciba		✓	✓	✓	Farm forestry label approval	Foliar	Hand-held and mechanised	0.7–1.5 l/ha	Maximum total applied to be 2.0 l/ha per year
Propyz-amide (400g/l, 50%w/w, 4%w/w)	Kerb Flowable/ Kerb 50W/ Kerb Granules***	P.B.I./ Rohm & Haas	✓	✓	✓	✓	Forestry label approval	Residual pre-emergent/ foliar	Hand-held and mechanised	3.75 l/ha 3.0 kg/ha 38.0 kg/ha	1
Simazine (500g/l)	Unicrop Flowable Simazine**	Unicrop				✓	LTOA – Short rotation coppice	Residual pre-emergent	Hand-held and mechanised	1.1–2.2 l/ha	1

Note:* The following additional glyphosate products have full on label approval for use in forestry and farm forestry:

Barbarian	360 g/litre glyphosate (Barclay)
Barclay Gallup	360 g/litre glyphosate (Barclay)
Barclay Gallup Amenity	360 g/litre glyphosate (Barclay)
Clayton Glyphosate	360 g/litre glyphosate (Clayton)
Clayton Swath	360 g/litre glyphosate (Clayton)
Glyfonex	360 g/litre glyphosate (Danagri)
Glyphos	360 g/litre glyphosate (Cheminova)
Glyphogan	360 g/litre glyphosate (PBI)
Glyphosate-360	360 g/litre glyphosate (Top Farm)
Helosate	360 g/litre glyphosate (Helm)
Hilite	144 g/litre glyphosate (Nomix-Chipman) – CDA formulation
Outlaw	360 g/litre glyphosate (Barclay)
Portman Glyphosate 360	360 g/litre glyphosate (Portman)
Roundup	360 g/litre glyphosate (Schering/AgrEvo)
Roundup	360 g/litre glyphosate (Monsanto)
Roundup Pro Biactive	360 g/litre glyphosate (Monsanto)
Roundup Biactive Dry	42.6% w/w glyphosate (Monsanto)
Stacato	360 g/litre glyphosate (Unicrop)
Stefes Glyphosate	360 g/litre glyphosate (Stefes)
Stetson	360 g/litre glyphosate (Monsanto)
Stefes Kickdown 2	360 g/litre glyphosate (Stefes)
Stirrup	144 g/litre glyphosate (Nomix-Chipman) – CDA formulation

These products may have different conditions of use - refer to the product label.

** The following additional products, with the same active ingredients as indicated in the Table, are approved for use under the long-term off-label arrangements for short rotation coppice. These products may have different conditions of use - refer to the product label.

cycloxidim	–	Stratos (BASF)
lenacil	–	Stefes lenacil (Stefes)
metamitron	–	Stefes 7G (Stefes)
		Stefes Metamitron (Stefes)
		Tripart Accendo (Tripart)
simazine	–	Ashlade Simazine (Ashlade)
		Atlas Simazine (Atlas)
		Gesatop 500SC (Ciba Agric)
		Gesatop 50WP (Ciba Agric)
		MSS Simazine 50FL (Mirfield)

*** The following products also have full label approval for use in forestry:

Atlas atrazine	500 g/litre atrazine (Atlas)
Atrazol	500g/litre atrazine (Sipcam)
Headland Judo	400g/litre propyzamide (Headland)
Headland Sword	150g/litre glufosinate ammonium (Headland)

agricultural sprayers. Plantations need to be designed to allow sprayer access, either through wide (2.8 metres plus) space between rows, or wider spacing at intervals to allow access of boom sprayers which can extend over rows of closer spaced trees. The use of treeshelters can make the application of non-selective herbicides easier. Further guidance on applicators is given in Forestry Commission Field Book 8 and Forestry Commission Technical Development Branch Technical Information Note 8/94.

Timing

General

Precise weeding regimes will depend on many factors such as crop species, weed species, site type, cultivation practice, etc. However a general regime for a lowland ex-arable site may be as follows:

1. Before any cultivation, clear any established weeds through the use of a broad spectrum contact herbicide.

2. If cultivation takes place, aim for a firm fine tilth for the effective use of residual herbicides.

3. Apply residual herbicides as overall or directed sprays immediately after planting to weed free sites.

4. During the growing season, repeat applications of residual/foliar or selective foliar acting herbicides to emerging weeds. In general, it is important to apply these products **to young weeds** before they become large and established. Alternatively, use directed sprays of broad spectrum herbicides to clear large established weeds.

5. Aim to clear up the site at the end of the growing season with applications of broad spectrum herbicides, if necessary directed away from crop species.

6. Repeat the regime (steps 3 to 5) in subsequent years until the crop trees are established and the dominant form of vegetation on the site, normally for a minimum of 3 years after planting.

Pre-emergent herbicides

Pre-emergent herbicides should be applied immediately after the trees have been planted, to weed-free sites prior to bud-burst. Subject to crop tolerance (see Table 3) most can be applied as an overall or directed spray. With poplar and willow cuttings, apply as soon as rain has consolidated soil around the cuttings (sets) as an overall or directed spray. If used in subsequent years, these herbicides should be applied to bare soil in early spring before weed emergence.

The propyzamide products Kerb 50W and Kerb Flowable should be applied from 1 October to 31 January, north of a line from Aberystwyth to London, and from 1 October to 31 December south of this line, and on peat or peaty gley soils. Kerb Granules should be applied from 1 October to the end of February, north of a line from Aberystwyth to London, and from 1 October to the end of January south of this line, and on peat or peaty gley soils.

Napropamide should be applied before the end of February, and isoxaben before the end of March.

For those products only approved for use after cut-back of short rotation coppice, apply prior to bud-burst, prior to weed germination, as an overall or directed spray.

If applied correctly these products may give weed control well into the growing season. Usually a mixture or sequence of products will be required, chosen according to the weed species present. In general, metamitron, metazachlor, atrazine and cyanazine have somewhat less effective residual properties than lenacil, napropamide, pendimethalin, simazine, isoxaben or propyzamide, having an effective life of about 12 weeks. Repeated applications of metazachlor and metamitron are permitted, but they should take place before weeds have passed the growth stage when they are susceptible.

For all these products it is important that rain follows application to move the herbicide into the top 2–3 cm of the soil. If these residual herbicides are applied to dry soil, and little or no rain follows application, weed control is likely to be poor.

Applications will be most effective when made to a firm, fine tilth. If the soil has large clods at the time of herbicide application, these may weather and crumble, exposing untreated soil and allowing prolific weed growth.

Foliar acting herbicides

The timing of application of foliar acting herbicides will be determined by the growth stage of the target weed (see Table 5). Applications may be made as overall or directed sprays. Dormant trees may be oversprayed by products such as fluazifop-p-butyl, cycloxidim and clopyralid. Overall sprays using these products may be safe when trees are actively growing (see Table 3) but it is advisable to avoid newly flushed trees, before new growth has hardened in the spring. When clopyralid is applied overall there may be some transient twisting of needles and young shoots, and cupping of leaves, but this will soon be outgrown.

Directed sprays of broad spectrum herbicides offer the least risk of crop damage and allow the use of higher product rates to control difficult weeds. However, in small-scale trials, sprays of glufosinate ammonium at 5 litres/ha have been shown to be safe for application over most broadleaved species (including willow and poplar cuttings), provided that the trees are deeply dormant. Glufosinate ammonium is at present (early 1996) only approved for use between 1 March and 30 September.

Glyphosate at 1.5 litres/ha can be used over dormant conifers. Results on broadleaves have been variable and it is advisable to use directed applications wherever possible.

Glyphosate is translocated more readily than glufosinate ammonium, so it will give better control of deeply rooted weed species. Conversely, accidental crop contamination from glufosinate ammonium through spray drift or applicators on to damaged bark is less likely to result in damage to the whole tree, and so is a safer treatment for in-season applications.

Amitrole should generally be used as a directed spray. However, in short rotation coppice overall applications of 20 litres/ha of product from 1 week after cutting, prior to bud burst, is well tolerated, although temporary yellowing of emerging shoots may occur. Applications made at higher rates are likely to reduce height growth and survival significantly, particularly with willow.

In general, applications of foliar acting herbicides should be avoided during periods of bright sunlight or high temperatures, as this can lead to scorching of tree foliage. If applications are necessary in mid-summer, they should be made in the evening, to allow the maximum delay between applications and the occurrence of bright sunlight and high temperatures. Rainfall shortly after spraying will seriously reduce the efficacy of all these products – consult the product label for details of the minimum rain free period required.

Applications of **ANY** herbicide to waterlogged ground, or to trees under stress from factors such as drought, should be avoided, as there is a greater risk of herbicide damage in these situations.

4 Crop tolerance

The herbicides listed in Table 1 are tolerated by the coniferous and broadleaved species listed in Table 2, when applied as detailed in Table 3. Information on products without full on-label approval is based upon small-scale research experiments. Users should determine the approval status of products before using them, and conduct their own limited field trials of new herbicides before adopting them on a commercial scale.

Table 2 Tree species screened for herbicide tolerance

Conifers	Broadleaves	Short rotation coppice
Sitka spruce	Oak	Poplar (sets)
Norway spruce	Ash	Willow (sets)
Douglas fir	Sycamore	
Noble fir	Beech	
Corsican pine	Wild cherry	
Western red cedar	Birch	
Japanese larch	Alder	
Scots pine	Sweet chestnut	
	Norway maple	
	Poplar (sets)*	
	Willow (sets)	

*A 'set' is a complete unrooted shoot. These were the stock type used in the poplar and willow herbicide screens.

Table 3 Crop tolerance

Active ingredient	Product	Trees dormant			Trees post-flushing		
		Conifers	Broadleaves	Short rotation coppice	Conifers	Broadleaves	Short rotation coppice
Amitrole	Weedazol TL	D	D	✓[4]	D	D	D
Atrazine	Unicrop Flowable Atrazine	✓[5]	✓[6]	✓[6]	✓[5]	X	X
Clopyralid	Dow Shield	✓	✓	✓	✓	✓	✓
Cyanazine	Fortrol	✓	✓	✓	✓	D	D
Cycloxidim	Laser	X	X	✓	X	X	✓
Fluazifop-p-butyl	Fusilade 250EW	✓	✓	✓	✓	✓	✓
Glufosinate ammonium	Challenge/Harvest	D	✓[7]	✓[7]	D	D	D
Glyphosate	Roundup Pro Biactive	✓[3]	D	D	D	D	D
Isoxaben	Gallery 125/ Flexidor 125	✓	✓	✓	✓	✓	✓
Lenacil	Venzar Flowable	X	X	✓	X	X	✓
Metamitron	Goltix WG	X	X	✓	X	X	D
Metazachlor	Butisan S	✓	✓	✓	✓[1]	✓	✓
Napropamide	Devrinol	X	X	✓	X	X	✓[2]
Pendimethalin	Stomp 400SC	✓	✓	✓	✓	✓	✓
Propaquizafop	Falcon/ Shogun 100EC	✓	✓	✓	✓	✓	✓
Propyzamide	Kerb Flowable/ Kerb 50W/ Kerb Granules	✓	✓	✓	✓[2]	✓[2]	✓[2]
Simazine	Unicrop Flowable Simazine	X	X	✓	X	X	✓

Notes:

- For the purposes of this table, treat larch as a broadleaved tree
- Trees will be at their most sensitive immediately after flushing. Herbicide application should not be made before new needles/leaves have hardened.
- In Forestry Commission trials, the treatments listed as safe to overspray trees post-flushing were found to have no significant effect on height or survival of the crop species listed. However, there may be some transient foliage damage. Where condition of foliage is particularly important, such as in Christmas trees, overall post-flushing applications are not recommended.

✓ = Herbicides can be used as an overall or directed spray.

D = Herbicides should only be used as a directed spray.

X = Herbicides are unapproved and must not be used.

✓[1] = In FC trials where metazachlor was applied to pine in active growth (e.g. candles fully extended but needles not fully hardened) damage was observed. The damage symptoms were distortion, browning and loss of needles from the tender new growth. On a few plants the growing tip was killed, but on most plants the main stem (or candle) remained healthy but devoid of needles. Terminal buds were set as normal.

✓[2] = By the time trees have flushed it is usually too late to achieve effective weed control using napropamide or propyzamide.

✓[3] = Sitka spruce, Norway spruce, Scots pine, Corsican pine, lodgepole pine, western red cedar and Lawson cypress will tolerate overall sprays at 1.5 l/ha PROVIDED TREES ARE DORMANT - i.e. STEM ELONGATION HAS CEASED, LEADER GROWTH HAS HARDENED, AND BUDS ARE TIGHTLY CLOSED.

✓[4] = See text. Overall spray after cutting, before regrowth, is well tolerated.

✓[5] = All the major forest species are tolerant of overall applications except Norway spruce, western hemlock and larch, which are sensitive during the growing season and should only be treated before bud burst.

✓[6] = All broadleaves are sensitive when in leaf, and should only be treated before the start of bud burst in the spring, at 6.5 l/ha. For both conifers and broadleaves, do not apply to badly planted trees, or those under stress, in poor health, or on light calcareous or sandy soils, or on reclaimed sites with poor soil structure.

✓[7] = Broadleaves will tolerate overall sprays at 5 l/ha PROVIDED TREES ARE DEEPLY DORMANT – i.e. STEM ELONGATION HAS CEASED, LEADER GROWTH HAS HARDENED, BUDS ARE TIGHTLY CLOSED, AND LEAVES HAVE BEEN SHED.

5 Weed susceptibility

The susceptibility of commonly occurring weeds to the herbicides listed in Table 1 are given in Tables 4 and 5. These tables are based on information supplied with product labels, and on limited Forestry Commission experience. In practice weeds may vary in susceptibility to a particular application for the reasons outlined in the earlier section on mode of action. Users are advised to make small-scale trials of products they are unfamiliar with before deciding to adopt them on a commercial scale.

Table 4 Susceptibility of common arable weeds to selective pre-emergent farm forestry herbicides

Weeds	Pre-emergent herbicides									
	atrazine	cyanazine	isoxaben	lenacil	metamitron	metazachlor	napropamide	pendimethalin	propyzamide	simazine
American willowherb				S	MS	S	S			
Bents	S								S	
Bitter cress, hairy			S	S	S	S	MS			
Bittersweet									S	
Black bindweed	MS	S		S	MR	MS	MS	S	S	MS
Black grass	S	MS		R	MR	S	S	MS	S	S
Black nightshade	S			MS			R	S	S	
Brome, barren		S				MS			S	
Buttercup, corn	R		S		MS			S		
Buttercup, creeping					MS				S	MR
Canary grass, awned								S		
Chamomile, corn	S		S	S		S		S		S
Chamomile, stinking	S		S	S		S		S		S
Charlock	S	S	S	S	MS	MR	R			S
Chickweed, common	S	S	S	S	S	S	S	S	S	S
Cleavers	MR	MR	MS	R	R	MS		S*	S	MR
Clover (from seed)					S					MR
Cocksfoot	MR								S	
Common couch	MR								S	
Crane's-bill, cut-leaved						S	S			MR
Creeping bent (watergrass)	S								S	
Creeping soft grass	MR		R						S	
Crested dog's tail									S	
Curled dock					S				S	

Table 4 Susceptibility of common arable weeds to selective pre-emergent farm forestry herbicides

Weeds	Pre-emergent herbicides									
	atrazine	cyanazine	isoxaben	lenacil	metamitron	metazachlor	napropamide	pendimethalin	propyzamide	simazine
Dead-nettle, henbit	S	S			MS		S	S		S
Dead-nettle, red	S	S	S	MS	MS	S	S	S		S
Dead-nettle, white	S	S						S		
Dock, broadleaved					S				S	
Established perennials		R								
False oat grass	MR								S	
Fat-hen	S	MS	S	S	S	MS	S	S	S	S
Fescues	MS								S	
Fescue, meadow	MS								S	
Field horsetail									S	
Fleabane, common						MS				
Fool's parsley		MR			S					
Forget-me-not, field	S	S	S		S	S		S	S	S
Foxglove									R	
Fumitory, common	MS	MS	S	S	MS	R	S	S	MS	MS
Gromwell, field						S				S
Groundsel	S	S	S	MS	S	S	S			S
Hemp-nettle, common	S	S	S	R		MR	S	S		S
Knotgrass	MR	MS	S	S	S	R	MS	S	S	MR
Marigold, corn	S		S	S	S	S	R	S		S
Mat grass									S	
Mayweed, scented	S	S	S	S	S	S	S	S		S
Mayweed, scentless	S	S	S	S	S	S	S	S		S
Meadow foxtail									S	
Meadow grass, annual	S	S	R	S	S	S	S	S	S	S
Meadow grass, rough	S	S	R					S	S	MR

5

Table 4 Susceptibility of common arable weeds to selective pre-emergent farm forestry herbicides

Weeds	Pre-emergent herbicides									
	atrazine	cyanazine	isoxaben	lenacil	metamitron	metazachlor	napropamide	pendimethalin	propyzamide	simazine
Meadow grass, smooth	MS		R						S	MR
Mustard, white	S	S				MR				S
Mustard, black	S	S				MR				S
Nettle, small	S	S	S	MS	S	MS	S	S	S	S
Nightshade, black	S	MS		R	MR			S	S	S
Orache, common	MS	S	S	S	S	S	S	S		MS
Pale persicaria	MS	S		S	MS		MS			MS
Pansy, field	MS	MS	S	R	S	MR	MS	S		MS
Pansy, wild						MR		S		
Parsley piert	S	S	S			S		S		S
Pimpernel, scarlet	S	S	S	S	MR		R	S		S
Pineapple weed	S		S	S		S	S	S		S
Poppy, common	S	S	S	S	S	S	S	S		S
Purple moor grass	R								S	
Radish, wild	S	S	S	S	MR					S
Redshank	MS	S	S	S	MS	MS	MS	S	S	MS
Rosebay willowherb	R								R	
Rye grasses	S	S					S		S	MS
Sedges									MS	
Sheep's sorrel	MR						S		S	
Shepherd's purse	S	S	S	S	S	S	MS	S	MS	S
Soft brome									S	
Speedwell, common	S	S	S	MR	S	S	S	S	S	MS
Speedwell, germander	S	S	S	MR		S		S		MR
Speedwell, ivy-leaved	S	S	S	R	MS	S		S	S	MS
Speedwell, grey	S	S	S	MR		S		S		MR

Table 4 Susceptibility of common arable weeds to selective pre-emergent farm forestry herbicides

Weeds	Pre-emergent herbicides									
	atrazine	cyanazine	isoxaben	lenacil	metamitron	metazachlor	napropamide	pendimethalin	propyzamide	simazine
Speedwell, green	S	S	S	MR		S		S		MR
Speedwell, wall	S	S	S			S		S		MS
Spurrey, corn	MS		S	S	S	MS	S		S	MS
Sweet vernal grass	S								S	
Timothy									S	
Thale cress	S					S				
Thistle, smooth sow				S			S	S		S
Tufted hair grass	MR								S	
Vetches (from seed)										MR
Volunteer cereals	MS					MR	MS		S	
Volunteer oilseed rape			S			R		S*		
Wavy hair grass	S								S	
Wild carrot					S					
Wild oat	MS			R	R	MR		S	S	MS
Wood small reed									S	
Yellow oat grass									S	
Yorkshire fog	S						S		S	

Key

S – susceptible
MS – moderately susceptible
MR – moderately resistant
R – resistant
– not tested
* – plants arising from deep-germinating seeds may not be controlled

5

Table 5 Susceptibility of common arable weeds to post-emergent farm forestry herbicides

Weeds	Post-emergent herbicides											
	amitrole°	atrazine	clopyralid	cyanazine	cycloxidim	fluazifop-p-butyl	glufosinate ammonium*	glyphosate§	metamitron	metazachlor	propaquizafop	propyzamide†
Bents	S	FT			FT	4ETL	S	S				S
Bitter cress, hairy	S						S	S				
Bittersweet	MS						MS	S				MS
Black bindweed	S	50mm	2ETL	100mm			S	S	MR			MS
Black grass	S	FT		2ETL	FT	FT	S	S	MR	2ETL	FT	S
Black nightshade	S	100mm					S	S				MS
Brome, barren	S			2ETL		FT	S	S				S
Buttercup, corn	S	R					S	S	MSC			
Buttercup, creeping	MS						S	S	MSC			MS
Canary grass, awned	MS						MS	S				
Chamomile, corn	S	100mm					S	S				
Chamomile, stinking	S	100mm	2ETL				S	S				
Charlock	S	100mm		6ETL			S	S	MSC			
Chickweed, common	S	100mm		100mm			S	S	C	4ETL		S
Cleavers	S						S	S	R			R
Clover (from seed)	S		2ETL				S	S	C			
Cocksfoot	S	MR					S	S				MR
Coltsfoot	S		MS6ETL#					S	S			
Common couch	S	3ETL			FT	4ETL	S	S			3ETL	S
Crane's-bill, cut-leaved	S						S	S		C		
Creeping bent (watergrass)	S	3ETL			FT	4ETL	MS	S				S
Creeping soft grass	S	MR					MS	S				MS
Crested dog's tail	S						S	S				S
Curled dock	S						S	S				MS
Dead-nettle, henbit	S	100mm		100mm			S	S	MSC			
Dead-nettle, red	S	100mm		100mm			S	S	MSC	2ETL		
Dead-nettle, white	S	100mm		100mm			S	S				

Table 5 Susceptibility of common arable weeds to post-emergent farm forestry herbicides

Weeds	Post-emergent herbicides										
	amitrole°	atrazine	clopyralid	cyanazine	cycloxidim	fluazifop-p-butyl	glufosinate ammonium*	glyphosate§	metamitron	metazachlor	propaquizafop	propyzamide†
Dock, broadleaved	S						S	S	C			MS
Established perennials	MS						MS	S				
False oat grass	S						S	S				S
Fat-hen	S	100mm		2ETL			S	S	MSC			MS
Fescues	S	MS					S	S				S
Fescue, meadow	S	MS					S	S				S
Field horsetail	MS						MS	MS				MS
Fleabane, common	S						S	S				
Fool's parsley	MS			1ETL			S	S	C			
Forget-me-not, field	S	100mm		4ETL			S	S	C	2ETL		S
Foxglove	MS						MS	S				R
Fumitory, common	S	50mm		1ETL			S	S	MSC			R
Gromwell, field	S						S	S				
Groundsel	S	100mm	6ETL	1ETL			S	S	MSC	2ETL		
Hemp-nettle, common	S			100mm			S	S				
Knotgrass	S		MS2ETL	1ETL			S	S	MSC			MS
Marigold, corn	S	100mm	6ETL				S	S	C	2ETL		
Mayweed, scented	S	100mm	6ETL	2ETL			S	S	C	4ETL		
Mayweed, scentless	S	100mm	6ETL	2ETL			S	S	C	4ETL		
Meadow foxtail	S						S	S				S
Meadow grass, annual	S	FT		FT	R		S	S	MSC	2ETL	3ETL	S
Meadow grass, rough	S	S		FT	MR		S	S				S
Meadow grass, smooth	S	MS					S	S				S
Mustard, white	S	100mm		6ETL			S	S				
Mustard, black	S	100mm		6ETL			S	S				
Nettle, small	S	100mm		100mm			S	S	C			MS
Nightshade, black	S	100mm		100mm			S	S	MR			MS

Table 5 Susceptibility of common arable weeds to post-emergent farm forestry herbicides

Weeds	amitrole[⊘]	atrazine	clopyralid	cyanazine	cycloxidim	fluazifop-p-butyl	glufosinate ammonium*	glyphosate[$]	metamitron	metazachlor	propaquizafop	propyzamide†
	Post-emergent herbicides											
Orache, common	S	50mm		1ETL			S	S	C			
Pale persicaria	S	50mm	MS2ETL	2ETL			S	S	MSC			
Pansy, field	S			1ETL			S	S	MSC			
Pansy, wild	S						S	S				
Parsley piert	S	100mm		1ETL			S	S				
Pimpernel, scarlet	S	100mm		100mm			S	S	MR			
Pineapple weed	S	100mm	6ETL				S	S		4ETL		
Poppy, common	S	100mm					S	S	C			
Purple moor grass	MS	R					MS	S				S
Radish, wild	S	50mm		100mm			S	S	MSC			
Redshank	S	50mm	MS2ETL	100mm			S	S	MSC			MS
Rosebay willowherb	MS	50mm					MS	S				R
Rye grasses	S	3ETL		3ETL		FT	S	S			FT	S
Sedges	MS						MS	S				MS
Sheep's sorrel	S						S	S				MS
Shepherd's purse	S	50mm		100mm			S	S	C			R
Soft brome	S						S	S				S
Speedwell, common	S	100mm		100mm			S	S	C	2ETL		MS
Speedwell, germander	S	100mm		100mm			S	S		2ETL		
Speedwell, ivy-leaved	S	100mm		100mm			S	S	MSC	2ETL		MS
Speedwell, grey	S	100mm		100mm			S	S		2ETL		
Speedwell, green	S	100mm		100mm			S	S		2ETL		
Speedwell, wall	S	100mm		100mm			S	S		2ETL		
Spurrey, corn	S	50mm					S	S	C			S
Sweet vernal grass	S	S					S	S				S
Timothy	S						S	S				S
Thale cress	MS						MS	S				

Table 5 Susceptibility of common arable weeds to post-emergent farm forestry herbicides

Weeds	Post-emergent herbicides: amitroleØ	atrazine	clopyralid	cyanazine	cycloxidim	fluazifop-p-butyl	glufosinate ammonium*	glyphosate$	metamitron	metazachlor	propaquizafop	propyzamide†
Thistle, creeping	S		MS250mm#				S	S				
Thistle, perennial sow	S		MS250mm#				S	S				
Thistle, smooth sow	S		6ETL				S	S				
Thistle, spear	MS		MS250mm#				MS	S				
Trefoils (from seed)	S		2ETL				S	S				
Tufted hair grass	S	MR					MS	S				S
Vetches (from seed)	MS		2ETL				S	S				
Volunteer cereals	S	FT			FT	FT	S	S			FT	S
Volunteer oilseed rape	S						S	S				
Wavy hair grass	S	S			FT		MS	S				S
Wild carrot	S		2ETL				S	S	C			
Wild oat	S	FT			FT	FT	S	S	R		FT	S
Wood small reed	MS						MS	S				S
Yellow oat grass	MS						S	S				S
Yorkshire fog	S	S					S	S				S

Key: *Post-emergent growth stage of weeds (latest at which controlled):*

C – cotyledon
ETL – number of expanded true leaves
ETLs – number of expanded true leaves (suppression only)
mm – diameter or height of weeds
Fbv – flower bud visible
FT – fully tillered
S – susceptible at all growth stages
MS – moderately susceptible at all growth stages
MR – moderately resistant at all growth stages
R – resistant
– not tested
† – all weed susceptibilities for propyzamide post emergence are for fully established weeds

Ø – amitrole will give a degree of control over most annual weeds present at application. Weed susceptibilities shown in this table are for fully established weeds.
– control with a programme of an application at 0.5 l/ha followed by one of 1.0 l/ha 3–4 weeks later
* – in addition to the weeds listed, most species will be damaged by applications of glufosinate ammonium at 5.0 l/ha if they are actively growing, although repeated applications may be required to achieve a total kill of deep rooted species
$ – in addition to the weeds listed, most species will be controlled by applications of glyphosate at 5.0 l/ha if they are actively growing.

5

6 Herbicide mixtures

Products which do not contain anti-cholinesterase ingredients (no anti-cholinesterase compounds are listed in this Field Book), can be used in tank mixes of two or more herbicides provided that all the conditions of use for all of the products to be used are complied with.

In agricultural situations, control of the wide range of weeds found on arable sites is commonly achieved using tank mixtures of herbicides. The application of tank mixes may be appropriate during tree establishment to cope with different mixtures of weeds and because of the absence of crop competition.

Results of trials carried out during 1989 and 1990 at a number of sites in southern Britain indicate that the herbicide mixtures listed in Table 6 are tolerated by the coniferous and broadleaved species listed in Table 2 when applied as overall sprays before bud-burst in spring. Propaquizafop was not subject to Forestry Commission trials - entries are based upon the manufacturer's data.

Table 6 Farm forestry herbicide tank mixes (all herbicides at approved rates)

Cyanazine	clopyralid isoxaben pendimethalin
Isoxaben	clopyralid cyanazine metazachlor propyzamide pendimethalin
Lenacil	metazachlor pendimethalin propyzamide
Metamitron	metazachlor
Metazachlor	clopyralid isoxaben lenacil metamitron pendimethalin propyzamide simazine
Pendimethalin	cyanazine isoxaben lenacil metazachlor propyzamide
Propaquizafop	clopyralid metazachlor
Propyzamide	clopyralid cyanazine isoxaben lenacil metazachlor pendimethalin
Simazine	metazachlor

Note: Mixtures of herbicides in the left column, with those listed to their right, were found to be safe. However, unless such mixes are specifically listed on the product label, they are made at the user's own risk.

Users MUST ALWAYS read the instructions on the herbicide product label, and follow the safety precautions and instructions therein relating to its use.

7 Further reading

Further detailed guidance on the use of herbicides can be found in:

WILLOUGHBY, I. and DEWAR, J. (1995). *Use of herbicides in the forest.* 4th edition. Forestry Commission Field Book 8. HMSO, London.

DRAKE-BROCKMAN, G.R. (1994). *An introduction to the use of tractor-mounted sprayers in farm woodland.* Technical Development Branch Information Note 8/94. Forestry Commission, Edinburgh. (Available only from: Technical Development Branch, Forestry Commission, Ae Village, Dumfries DG1 1QB).

8 Appendix

Notices of approval

NOTICE OF APPROVAL No. 0757/92

FOOD AND ENVIRONMENT PROTECTION ACT 1985
CONTROL OF PESTICIDES REGULATIONS 1986
(S.I. 1986 No. 1510):
APPROVAL FOR OFF-LABEL USE OF AN APPROVED PESTICIDE PRODUCT

This approval provides for the use of the product named below in respect of crops and situations, other than those included on the product label. Such 'off-label use' as it is known is at all times done at the user's choosing, and the commercial risk is entirely his or hers.

The conditions below are statutory. They must be complied with when the off-label use occurs. Failure to abide by the conditions of approval may constitute a breach of that approval, and a contravention of the Control of Pesticides Regulations 1986. The conditions shown below supersede any on the label *which would otherwise apply.*

Level and scope:	In exercise of the powers conferred by regulation 5 of the Control of Pesticides Regulations 1986 (SI 1986/1510) and of all other powers enabling them in that behalf, the Minister of Agriculture, Fisheries and Food and the Secretary of State, hereby jointly give full approval for the advertisement, sale, supply, storage and use of
Product name:	Dow Shield containing

8

Active ingredient:	200 g/l clopyralid
Marketed by:	DowElanco Ltd under MAFF No. 05578 subject to the conditions relating to off-label use set out below:
Date of issue:	15 July 1992
Date of expiry:	(unlimited subject to the continuing approval of MAFF 05578)
Field of use:	ONLY AS A FORESTRY HERBICIDE
Crops:	Coniferous and broadleaved trees
Maximum individual dose:	1 litre product/hectare
Maximum number of treatments:	Two per year
Operator protection:	(1) Engineering control of operator exposure must be used where reasonably practicable in addition to the following personal protective equipment: Operators must wear suitable protective gloves and face protection (faceshield) when handling the concentrate. (2) However, engineering controls may replace personal protective equipment if a COSHH assessment shows they provide an equal or higher standard or protection.

Other specific restrictions:

(1) This product must only be applied if the terms of this approval, the product label and/or leaflet and any additional guidance on off-label approvals have first been read and understood.

(2) Livestock must be kept out of treated areas for at least 7 days following treatment and until poisonous weeds such as ragwort have died and become unpalatable.

Signed J Micklewright
(Authorised signatory)

Date 15 July 1992

Application Reference Number: COP 91/00752

THIS NOTICE OF APPROVAL IS NUMBER 0757 of 1992

ADVISORY INFORMATION

This approval relates to the use of Dow Shield on coniferous and broadleaved trees. Overall spray may result in crop damage. Tree guards should be used to prevent foliar contamination. When spraying with a knapsack sprayer use 1 part product to 240 parts water. Do not spray to run-off.

NOTICE OF APPROVAL No. 0226/94

FOOD AND ENVIRONMENT PROTECTION ACT 1985
CONTROL OF PESTICIDES REGULATIONS 1986
(S.I. 1986 No. 1510):
APPROVAL FOR OFF-LABEL USE OF AN APPROVED PESTICIDE PRODUCT

This approval provides for the use of the product named below in respect of crops and situations, other than those included on the product label. Such 'off-label use', as it is known, is at all times done at the user's choosing, and the commercial risk is entirely his or hers.

The conditions below are statutory. They must be complied with when the off-label use occurs. Failure to abide by the conditions of approval may constitute a breach of that approval, and a contravention of the Control of Pesticides Regulations 1986. The conditions shown below supersede any on the label *which would otherwise apply.*

Level and scope: In exercise of the powers conferred by regulation 5 of the Control of Pesticides Regulations 1986 (SI 1986/1510) and of all other powers enabling them in that behalf, the Minister of Agriculture, Fisheries and Food and the Secretary of State, hereby jointly give full approval for the use of

Product name: Stomp 400 SC containing

Active ingredient: 400 g/l pendimethalin

Marketed by: Cyanamid of Great Britain Ltd under MAFF No. 04183 subject to the conditions relating to off-label use set out below:

Date of issue:	27 January 1994
Date of expiry:	Unlimited subject to the continuing approval of MAFF 04183
Field of use:	ONLY AS AN AGRICULTURAL/ FORESTRY HERBICIDE
Situations:	Farm forestry
Maximum individual dose:	5 litres product/hectare
Maximum number of treatments:	One per year
Environmental protection:	Since this product is dangerous to fish or aquatic life, surface waters or ditches must not be contaminated with chemical or used container.
Other specific restrictions:	(1) This product must only be applied if the terms of this approval, the product label and/or leaflet and any additional guidance on off-label approvals have first been read and understood. (2) Crops grown within the treated area of woodland must not be used for human or animal consumption. (3) This product must only be used for forestry establishment on land previously under arable

8

cultivation or improved grassland.

Signed J Micklewright
(Authorised signatory)

Date 27 January 1994

Application Reference Number: COP 93/01112

THIS NOTICE OF APPROVAL IS NUMBER 0226 of 1994

THIS NOTICE OF APPROVAL SUPERSEDES NOTICE OF APPROVAL No. 0377 OF 1991.

ADVISORY INFORMATION
This approval relates to the use of Stomp 400 SC in areas of conversion from arable land or improved grassland to farm forestry and coppice woodland at a maximum rate of 5 litres product/hectare applied in 200-400 litres water/hectare via tractor-mounted or knapsack sprayer.

NOTICE OF APPROVAL No. 0227/94

FOOD AND ENVIRONMENT PROTECTION ACT 1985
CONTROL OF PESTICIDES REGULATIONS 1986
(S.I. 1986 No. 1510):
APPROVAL FOR OFF-LABEL USE OF AN APPROVED PESTICIDE PRODUCT

This approval provides for the use of the product named below in respect of crops and situations, other than those included on the product label. Such 'off-label use', as it is known, is at all times done at the user's choosing, and the commercial risk is entirely his or hers.

The conditions below are statutory. They must be complied with when the off-label use occurs. Failure to abide by the conditions of approval may constitute a breach of that approval, and a contravention of the Control of Pesticides Regulations 1986. The conditions shown below supersede any on the label *which would otherwise apply.*

8

Level and scope:	In exercise of the powers conferred by regulation 5 of the Control of Pesticides Regulations 1986 (SI 1986/1510) and of all other powers enabling them in that behalf, the Minister of Agriculture, Fisheries and Food and the Secretary of State, hereby jointly give full approval for the use of
Product name:	Butisan S containing
Active ingredient:	500 g/l metazachlor
Marketed by:	BASF plc under under MAFF No. 00357 subject to the conditions relating to off-label use set out below:
Date of issue:	27 January 1994

Date of expiry:	Unlimited subject to the continuing approval of MAFF 00357
Field of use:	ONLY AS AN AGRICULTURAL/FORESTRY HERBICIDE
Situations:	Farm forestry
Maximum individual dose:	2.5 litres product/hectare
Maximum number of treatments:	Three per year
Operator protection:	(1) Engineering control of operator exposure must be used where reasonably practicable in addition to the following personal protective equipment: (a) Operators must wear suitable protective gloves and face protection (faceshield) when handling the concentrate. (b) Operators must wear suitable protective gloves when handling the spray boom or adjusting nozzles. (2) However, engineering controls may replace personal protective equipment if a COSHH assessment shows they provide an equal or higher standard of protection.

Environmental protection:	Since this product is dangerous to fish or aquatic life, surface waters or ditches must not be contaminated with chemical or used container.
Other specific restrictions:	(1) This product must only be applied if the terms of this approval, the product label and/or leaflet and any additional guidance on off-label approvals have first been read and understood. (2) Livestock must be kept out of treated areas until foliage of any poisonous weeds, such as ragwort, have died and become unpalatable. (3) Crops grown within the treated area of woodland must not be used for human or animal consumption. (4) This product must only be used for forestry establishment on land previously under arable cultivation or improved grassland.

Signed J Micklewright
(Authorised signatory)

Date 27 January 1994

Application Reference Number: COP 93/01118

THIS NOTICE OF APPROVAL IS NUMBER 0227 of 1994

THIS NOTICE OF APPROVAL SUPERSEDES NOTICE OF APPROVAL No. 0191 OF 1990.

ADVISORY INFORMATION

This approval relates to the use of 'Butisan S' in areas of conversion from arable land or improved grassland to farm forestry and coppice woodland. Application may be made by converntional tractor-mounted ground spray equipment or knapsack sprayers.

NOTICE OF APPROVAL No. 0602/94

FOOD AND ENVIRONMENT PROTECTION ACT 1985
CONTROL OF PESTICIDES REGULATIONS 1986
(S.I. 1986 No. 1510):
APPROVAL FOR OFF-LABEL USE OF AN APPROVED PESTICIDE PRODUCT

This approval provides for the use of the product named below in respect of crops and situations, other than those included on the product label. Such 'off-label use', as it is known, is at all times done at the user's choosing, and the commercial risk is entirely his or hers.

The conditions below are statutory. They must be complied with when the off-label use occurs. Failure to abide by the conditions of approval may constitute a breach of that approval, and a contravention of the Control of Pesticides Regulations 1986. The conditions shown below supersede any on the label *which would otherwise apply.*

Level and scope:	In exercise of the powers conferred by regulation 5 of the Control of Pesticides Regulations 1986 (SI 1986/1510) and of all other powers enabling them in that behalf, the Minister of Agriculture, Fisheries and Food and the Secretary of State, hereby jointly give full approval for the use of
Product name:	Fortrol containing
Active ingredient:	500 g/l cyanazine
Marketed by:	Cyanamid UK Limited under MAFF No. 07009 subject to the conditions relating to off-label use set out below:
Date of issue:	6 April 1994

8

Date of expiry:	Unlimited (subject to the continuing approval of MAFF 07009)
Field of use:	ONLY AS A FORESTRY HERBICIDE
Situations:	Farm forestry
Maximum individual dose:	4 litres product/hectare
Maximum number of treatments:	One per year
Operator protection:	(1) Engineering control of operator exposure must be fused where reasonably practicable in addition to the following personal protective equipment: Operators must wear suitable protective clothing (coveralls), suitable protective gloves and face protection (faceshield) when handling the concentrate. (2) However, engineering controls may replace personal protective equipment if a COSHH assessment shows they provide an equal or higher standard of protection.
Environmental protection:	Since this product is harmful to fish or aquatic life, surface waters or ditches must not be contaminated with chemical or used container.
Other specific restrictions:	(1) This product must only be applied if the terms of this

approval, the product label and/or leaflet and any additional guidance on off-label approvals have first been read and understood.

(2) Crops grown within the treated area of woodland must not be used for human or animal consumption.

(3) This product must only be used for forestry establishment on land previously under arable cultivation or improved grassland.

8

Signed J Micklewright
(Authorised signatory)

Date 6 April 1994

Application Reference Number: COP 93/01119

THIS NOTICE OF APPROVAL IS NUMBER 0602 OF 1994

THIS NOTICE OF APPROVAL SUPERSEDES NOTICE OF APPROVAL No. 0192 OF 1991

ADVISORY INFORMATION
This approval relates to the use of 'Fortrol' in areas of conversion from arable land or improved grassland to farm forestry and coppice woodland, to be applied via conventional ground-based machinery (vehicle-mounted hydraulic sprayers and knapsack sprayers).
Please note that an identical approval has been issued for this off-label use for 'Fortrol', marketed by Shell Chemicals UK Ltd, MAFF 00924.

NOTICE OF APPROVAL No. 0603/94

FOOD AND ENVIRONMENT PROTECTION ACT 1985
CONTROL OF PESTICIDES REGULATIONS 1986
(S.I. 1986 No. 1510):
APPROVAL FOR OFF-LABEL USE OF AN APPROVED PESTICIDE PRODUCT

This approval provides for the use of the product named below in respect of crops and situations, other than those included on the product label. Such 'off-label use', as it is known, is at all times done at the user's choosing, and the commercial risk is entirely his or hers.

The conditions below are statutory. They must be complied with when the off-label use occurs. Failure to abide by the conditions of approval may constitute a breach of that approval, and a contravention of the Control of Pesticides Regulations 1986. The conditions shown below supersede any on the label *which would otherwise apply.*

Level and scope:	In exercise of the powers conferred by regulation 5 of the Control of Pesticides Regulations 1986 (SI 1986/1510) and of all other powers enabling them in that behalf, the Minister of Agriculture, Fisheries and Food and the Secretary of State, hereby jointly give full approval for the use of
Product name:	Fortrol containing
Active ingredient:	500 g/l cyanazine
Marketed by:	Shell Chemicals UK Ltd under MAFF 00924 subject to the conditions relating to off-label use set out below:
Date of issue:	6 April 1994

Date of expiry:	Unlimited (subject to the continuing approval of MAFF 00924)
Field of use:	ONLY AS A FORESTRY HERBICIDE
Situations:	Farm forestry
Maximum individual dose:	4 litres product/hectare
Maximum number of treatments:	One per year
Operator protection:	(1) Engineering control of operator exposure must be fused where reasonably practicable in addition to the following personal protective equipment: Operators must wear suitable protective clothing (coveralls), suitable protective gloves and face protection (faceshield) when handling the concentrate. (2) However, engineering controls may replace personal protective equipment if a COSHH assessment shows they provide an equal or higher standard of protection.
Environmental protection:	Since this product is harmful to fish or aquatic life, surface waters or ditches must not be contaminated with chemical or used container.

Other specific restrictions:

(1) This product must only be applied if the terms of this approval, the product label and/or leaflet and any additional guidance on off-label approvals have first been read and understood.

(2) Crops grown within the treated area of woodland must not be used for human or animal consumption.

(3) This product must only be used for forestry establishment on land previously under arable cultivation or improved grassland.

Signed J Micklewright
(Authorised signatory)

Date 6 April 1994

Application Reference Number: COP 93/01119

THIS NOTICE OF APPROVAL IS NUMBER 0603 OF 1994

THIS NOTICE OF APPROVAL SUPERSEDES NOTICE OF APPROVAL No. 0192 OF 1991

ADVISORY INFORMATION
This approval relates to the use of 'Fortrol' in areas of conversion from arable land or improved grassland to farm forestry and coppice woodland, to be applied via conventional ground-based machinery (vehicle-mounted hydraulic sprayers and knapsack sprayers).
Please note that an identical approval has been issued for this off-label use for 'Fortrol', marketed by Cyanamid UK Ltd, MAFF 07009.

NOTICE OF APPROVAL No. 1302/94

FOOD AND ENVIRONMENT PROTECTION ACT 1985
CONTROL OF PESTICIDES REGULATIONS 1986
(S.I. 1986 NO. 1510):
APPROVAL FOR OFF-LABEL USE OF AN APPROVED PESTICIDE PRODUCT

This approval provides for the use of the product named below in respect of crops and situations, other than those included on the product label. Such 'off-label use', as it is known, is at all times done at the user's choosing, and the commercial risk is entirely his or hers.

The conditions below are statutory. They must be complied with when the off-label use occurs. Failure to abide by the conditions of approval may constitute a breach of that approval, and a contravention of the Control of Pesticides Regulations 1986. The conditions shown below supersede any on the label *which would otherwise apply.*

8

Level and scope:	In exercise of the powers conferred by regulation 5 of the Control of Pesticides Regulations 1986 (SI 1986/1510) and of all other powers enabling them in that behalf, the Minister of Agriculture, Fisheries and Food and the Secretary of State hereby jointly give full approval for the use of
Product name:	Fusilade 250 EW containing
Active ingredient:	250 g/l fluazifop-p-butyl
Marketed by:	Zeneca Crop Protection under MAFF 06531 subject to the conditions relating to off-label use set out below:
Date of issue:	9 August 1994

Date of expiry:	Unlimited, subject to the continuing approval of MAFF 06531.
Field of use:	ONLY AS AN AGRICULTURAL/ HORTICULTURAL/ FORESTRY HERBICIDE
Situations:	Farm forestry
Maximum individual dose:	1.5 litres product/hectare
Maximum number of treatments:	Two per year
Operator protection:	(1) Engineering control of operator exposure must be used where reasonably practicable in addition to the following personal protective equipment: (a) Operators must wear suitable protective clothing (coveralls) suitable protective gloves and face protection (faceshield) when handling the concentrate. (b) Operators must wear suitable protective clothing (coveralls) and suitable protective gloves when spraying and adjusting or maintaining equipment or handling contaminated surfaces.

	(2) However, engineering controls may replace personal protective equipment if a COSHH assessment shows they provide an equal or higher standard of protection.
Environmental protection:	Since this product is harmful to fish or aquatic life, surface waters or ditches must not be contaminated with chemical or used container.
Other specific restrictions:	(1) This product must only be applied if the terms of this approval, the product label and/or leaflet and any additional guidance on off-label approvals have first been read and understood.
	(2) This product must not be applied by hand-held equipment.
	(3) This product must only be used on land previously under arable cultivation or improved grassland.

Signed J Micklewright
(Authorised signatory)

Date 24 August 1994

Application Reference Number: COP 94/00589

THIS NOTICE OF APPROVAL IS NUMBER 1302 of 1994

ADVISORY INFORMATION

This approval relates to the use of 'Fusilade 250 EW' in areas of conversion from arable land or improved grassland to farm forestry and coppice woodland at a maximum rate of 1.5 litres product/hectare, applied in 200-500 litres water/hectare via tractor-mounted sprayers only.

8

NOTICE OF APPROVAL No. /94

FOOD AND ENVIRONMENT PROTECTION ACT 1985
CONTROL OF PESTICIDES REGULATIONS 1986
(S.I. 1986 NO. 1510):
APPROVAL FOR OFF-LABEL USE OF AN APPROVED PESTICIDE PRODUCT

This approval provides for the use of the product named below in respect of crops and situations, other than those included on the product label. Such 'off-label use', as it is known, is at all times done at the user's choosing, and the commercial risk is entirely his or hers.

The conditions below are statutory. They must be compiled with when the off-label use occurs. Failure to abide by the conditions of approval may constitute a breach of that approval, and a contravention of the Control of Pesticides Regulations 1986. The conditions shown below supersede any on the label *which would otherwise apply.*

Level and scope: In exercise of the powers conferred by regulation 5 of the Control of Pesticides Regulations 1986 (SI 1986/1510) and of all other powers enabling them in that behalf, the Minister of Agriculture, Fisheries and Food and the Secretary of State, hereby jointly give full approval for the use of

Product name: Unicrop Flowable Atrazine containing

Active ingredient: 500 g/l atrazine

Marketed by: Universal Crop Protection Ltd under MAFF 02268 subject to the conditions relating to off-label use set out below:

Date of issue:	16 August 1994
Date of expiry:	Unlimited, subject to the continuing approval of MAFF 02268
Field of use:	ONLY AS A FORESTRY HERBICIDE
Crop/situation:	Forestry (broadleaf)
Maximum individual dose:	13.5 litres product/hectare
Maximum number of treatments:	One per year
Operator protection:	(1) Engineering control of operator exposure must be used where reasonably practicable in addition to the following personal protective equipment:
	(a) Operators must wear suitable protective clothing (coveralls) and suitable protective gloves when handling contaminates surfaces.
	(b) Operators must wear suitable protective clothing (coveralls) when applying by vehicle-mounted equipment.
	(c) Operators must wear suitable protective clothing (coveralls), suitable protective gloves, boots, face protection (faceshield) and

8

respiratory/protective equipment (disposable filtering facepiece respirator) when applying by hand-held equipment.

(d) Operators must wear suitable protective clothing (coveralls), suitable protective gloves and face protection (faceshield) when handling the concentrate.

(2) However, engineering controls may replace personal protective equipment if a COSHH assessment shows they provide an equal or higher standard of protection.

Environmental protection:

(1) Since this product is dangerous to fish or aquatic life and aquatic higher plants, surface waters or ditches must not be contaminated with chemical or used container.

(2) Since there is a risk to aquatic life from use, direct spray from ground-based vehicle-mounted/drawn sprayers must not be allowed to fall within 6 m of surface waters or ditches; direct

spray from hand-held sprayers must not be allowed to fall within 2 m of surface waters or ditches; spray must be directed away from water.

Other specific restrictions:

(1) This product must only be applied if the terms of this approval, the product label and/or leaflet and any additional guidance on off-label approvals have first been read and understood.

(2) Use must be restricted to one product containing atrazine or simazine either to a single application at the maximum approved rate of (subject to any existing maximum permitted number of treatments) to several applications at lower doses up to the maximum approved rate for a single application.

Signed J Micklewright
(Authorised signatory)

Date 16 August 1994

Application Reference Number: COP 87/00873

THIS NOTICE OF APPROVAL IS NUMBER 1338 of 1994

ADVISORY INFORMATION
This approval relates to the use of 'Unicrop Flowable Atrazine' as a herbicide for use in broadleaved forestry or woodland, to control grass and herbaceous weeds. The product is applied from February to May, in a minimum of 200 litres water/hectare, by means of tractor-mounted or hand-held sprayers.
Please note that an identical off-label approval has been issued for MAFF 05446.

NOTICE OF APPROVAL No. /94

FOOD AND ENVIRONMENT PROTECTION ACT 1985
CONTROL OF PESTICIDES REGULATIONS 1986
(S.I. 1986 NO. 1510):
APPROVAL FOR OFF-LABEL USE OF AN APPROVED PESTICIDE PRODUCT

This approval provides for the use of the product named below in respect of crops and situations, other than those included on the product label. Such 'off-label use', as it is known, is at all times done at the user's choosing, and the commercial risk is entirely his or hers.

The conditions below are statutory. They must be compiled with when the off-label use occurs. Failure to abide by the conditions of approval may constitute a breach of that approval, and a contravention of the Control of Pesticides Regulations 1986. The conditions shown below supersede any on the label *which would otherwise apply.*

Level and scope:	In exercise of the powers conferred by regulation 5 of the Control of Pesticides Regulations 1986 (SI 1986/1510) and of all other powers enabling them in that behalf, the Minister of Agriculture, Fisheries and Food and the Secretary of State, hereby jointly give full approval for the use of
Product name:	Unicrop Flowable Atrazine containing
Active ingredient:	500 g/l atrazine
Marketed by:	Universal Crop Protection Ltd under MAFF 05446 subject to

8

	the conditions relating to off-label use set out below:
Date of issue:	16 August 1994
Date of expiry:	Unlimited, subject to the continuing approval of MAFF 05446
Field of use:	ONLY AS A FORESTRY HERBICIDE
Crop/situation:	Forestry (broadleaf)
Maximum individual dose:	13.5 litres product/hectare
Maximum number of treatments:	One per year
Operator protection:	(1) Engineering control of operator exposure must be used where reasonably practicable in addition to the following personal protective equipment: (a) Operators must wear suitable protective clothing (coveralls) and suitable protective gloves when handling contaminates surfaces. (b) Operators must wear suitable protective clothing (coveralls) when applying by vehicle-mounted equipment. (c) Operators must wear suitable protective clothing (coveralls), suitable protective gloves, boots, face

protection (faceshield) and respiratory/ protective equipment (disposable filtering facepiece respirator) when applying by hand-held equipment.

(d) Operators must wear suitable protective clothing (coveralls), suitable protective gloves and face protection (faceshield) when handling the concentrate.

(2) However, engineering controls may replace personal protective equipment if a COSHH assessment shows they provide an equal or higher standard of protection.

Environmental protection:

(1) Since this product is dangerous to fish or aquatic life and aquatic higher plants, surface waters or ditches must not be contaminated with chemical or used container.

(2) Since there is a risk to aquatic life from use, direct spray from ground-based vehicle-mounted/drawn sprayers must not be allowed to fall within 6 m of surface

		waters or ditches; direct spray from hand-held sprayers must not be allowed to fall within 2 m of surface waters or ditches; spray must be directed away from water.
Other specific restrictions:	(1)	This product must only be applied if the terms of this approval, the product label and/or leaflet and any additional guidance on off-label approvals have first been read and understood.
	(2)	Use must be restricted to one product containing atrazine or simazine either to a single application at the maximum approved rate of (subject to any existing maximum permitted number of treatments) to several applications at lower doses up to the maximum approved rate for a single application.

Signed J Micklewright
(Authorised signatory)

Date 16 August 1994

Application Reference Number: COP 87/00873

THIS NOTICE OF APPROVAL IS NUMBER 1339 of 1994

ADVISORY INFORMATION
This approval relates to the use of 'Unicrop Flowable Atrazine' as a herbicide for use in broadleaved forestry or woodland, to control grass and herbaceous weeds. The product is applied from February to May, in a minimum of 200 litres water/hectare, by means of tractor-mounted or hand-held sprayers.
Please note that an identical off-label approval has been issued for MAFF 02268.

Long-term off-label arrangements

The long-term arrangements confer off-label approval to certain fields of use, rather than to specific products. The same basic principles as specific off-label approval apply, namely users must comply with all label conditions of use as well as additional off-label restrictions, and all applications are made at the user's own risk.

Fields of use

The following extensions of use are permitted under the long-term arrangements:

- Herbicides with full or provisional label approval for use on cereals, may be used in the first five years of establishment of new farm woodlands (including short rotation energy coppice), on land previously under arable cultivation or improved grassland (as defined in the Woodland Grant Scheme).
- Herbicides with full or provisional label approval for use on cereals, oil-seed rape, sugar beet, potatoes, peas and beans, may be used in the first year of regrowth following cutting in short rotation energy coppice, on land previously under cultivation or improved grassland (as defined in the Woodland Grant Scheme).

Conditions of use

As well as the usual good working practices required of users, the following additional conditions MUST be complied with when applying pesticides under the long-term off-label arrangements.

- All precautions and statutory conditions of use which are identified on the product label, must be observed.
- The method of application used must be the same as that listed on the product label, and comply with relevant codes of practice and requirements under COSHH.
- All reasonable precautions must be taken to safeguard wildlife and the environment.

- Products must not be used in or near water unless the label specifically permits this.
- Aerial applications are not permitted.
- Products approved for use under protection, i.e. under polythene tunnels or glasshouses, cannot be used outside.
- Rodenticides and other vertebrate control agents are not included in these arrangements.
- Use is not permitted on land not intended for cropping, e.g. paths, roads, around buildings, wild mountainous areas, nature reserves, etc.
- Pesticides classified as hazardous to bees must not be applied when crops or weeds are flowering.
- These extensions of use apply only to label recommendations – no extrapolations are permitted from specific off-label approvals.
- Unless specifically permitted on the product label, hand-held applications are NOT permitted if the product label:

– prohibits hand-held use

– requires the use of personal protective clothing when using the pesticide at recommended volume rates

– is classified as 'corrosive', 'very toxic', 'toxic', or has a 'risk of serious damage to eyes'.

If none of the above apply hand-held application IS permitted provided that:

– the concentration of the spray volume does not exceed the maximum recommended on the label;

– spray quality is at least as coarse as the British Crop Protection Council medium or coarse spray;

– operators wear a protective coverall, boots and gloves for applications below waist height, and in addition a face shield is worn for applications above waist height;

– if the product gives a buffer zone for vehicle-mounted use, a buffer zone of 2 metres should be used for hand-held applications.

Printed in the United Kingdom for HMSO.
Dd 301936, 4/96, C17, 3400, 5673, 344875.